BIRDS & BEES

*Ectophylla alba*

(Actual size)

# BIRDS & BEES

## A SEXUAL STUDY

TEXT

&

ILLUSTRATIONS

BY

DUGALD

STERMER

CollinsPublishersSanFrancisco

*A Division of* HarperCollins*Publishers*

First published in USA in 1995 by
Collins Publishers San Francisco
1160 Battery Street, San Francisco CA 94111

In collaboration with Flying Fish Books and
Swans Island Books

Library of Congress Cataloging-in-Publication Data

Stermer, Dugald, 1936—
    Birds & Bees : a sexual study / Dugald Stermer.
        p.    cm.
    ISBN 0-00-255462-3
    1. Sexual behavior in animals. 2. Sexual behavior
in animals—Pictorial works. 3. Parental behavior in
animals.    4. Parental behavior in animals—Pictorial
works.            I. Title.
    QL761.S67   1995
    591.1'6—dc20                        95-10192

Printed in Italy

10 9 8 7 6 5 4 3 2 1

To Mary Blue Stermer

For Dugald II, Bridget, Megan, Sam, Chris, Colin, May and Crystal

# CONTENTS

# INTRODUCTION

*Detailed theoretical analysis, using mathematical modeling, reveals that sex is a very inefficient way of reproducing. The inefficiency lies in making male offspring. In the majority of sexual species only the female contributes energy and resources to the young. In contrast the males rarely contribute more than the minimum—a tiny sperm carrying genes, but devoid of other resources.*

—*The Evolution of Life*, edited by Linda Gamlin and Gail Vines,
Oxford University Press, New York, 1987

IF NATURE has forced this inefficiency on the female, at least in many of the world's species she chooses whose genes will pair with her own in the drive to perpetuate and perhaps even to improve the race. If gestation and rearing can be exhausting for the female, the overwhelming impulse to mate is often daunting, dangerous, and even fatal for the male.

Q: *You said that the male lovebug on the average devotes 56 hours of its life to making love. How long does it live?*

A: *Only a little more than 56 hours, understandably.*
—From *The Grab Bag* newspaper column, by L.M. Boyd

To attract a female, the male preens, dances, calls, poses, fights off other suitors, builds structures, brings gifts, sings, grooms, and performs daredevil feats. His reward can be, in addition to passing on his genes to future generations, injury and death. The female Chinese mantis bites off the head of her partner during coitus; afterward, instinct propels him into finishing the act. The male marsupial mouse of Australia scampers around frantically, mating repeatedly with several females for up to twelve hours, until he collapses and dies from exhaustion. The female redback spider, far larger than the doomed male, devours him after mating. However, before she does, he inserts a plug into her, a chastity belt, insuring that his will be the only genes to be passed on.

In a few species the males become female after mating, or attach themselves to the nearest female soon after birth, then to live out their lives as diminutive sexual appendages.

On the other hand, before mating the male porcupine urinates on his intended.

Most of nature, however, is not so disrespectful in its quest to continue. In fact, we *Homo sapiens* might identify with many of the antics performed in the service of procreation, and envy others. For many years, observers of the coneheaded katydid were stumped as to why the males, unlike other similar insects, sang in unison instead of solo. It was later discovered that the female would only mate with the male who sang one-tenth of a second faster than any of his rivals. Male bowerbirds build elaborate structures to attract a mate. If their efforts are successful, this bower is used once for copulation, then abandoned. And bonobo chimpanzees spend their entire lives in sexual play. Other than sleeping and eating, it's about all they ever do.

Presenting gifts, especially food, is, for the males of many species, including our own, a preferred method for attracting a mate. Such gift giving not only demonstrates skill in hunting and a predisposition toward sharing, but also indicates—often falsely—that the male might be useful in the rearing process as well.

The male shrike, a smallish bird, has developed a particularly graphic method for displaying his prowess as a hunter; he impales his prey—insects, snakes, other birds, lizards, grasshoppers, beetles, and even small mammals—upon thorns, sharp branches, barbed wire, and any other suitable, highly visible, spines. Close research and testing has indicated that because the male strews about his territory far more carcasses than he can eat, especially just prior to the breeding season, this grisly exhibit can only be meant to impress incoming female shrikes. Males with the most lavish larders nested earlier, more often, and with more mates, than did their less well provisioned brethren. Those with nothing on their pikestaffs remained bachelors, soon leaving the area forever.

As has been well documented and analogized, size plays an enormous role in sexual politics, at least in the natural world. The bigger the bull, the more likely it is that he will be able, through battle or intimidation, to exclude all other males in the herd from genetic immortality. It is also generally true that the larger the male in relation to the female of the species, the more of a harem he is able to control. On the other side of that coin, consider the bonellia worm, a little-known marine creature. Females are roughly a thousand times larger than the males, many of whom live in their partner's genital tract where they fertilize her eggs.

Finally, in nature's battle of the sexes, there are winners and losers, these latter being those whose genes are discontinued with their death. But it's never for lack of trying. Sex is an imperative, an irresistible drive that can only be thwarted by the refusal of available partners, by losing in competition with other hopefuls, or by abject incompetence. Abstinence or chastity is never an option, except in humans. Well, some humans. To my knowledge, we are the only organisms that can choose both who *and* whether. It is probably too early in our history to assess if that is an evolutionary advance or yet another experimental misstep.

I began this book out of curiosity, which has now turned into awe. I hope a little of it rubs off on the reader.

FOREPLAY

*Taeniopyga guttata*

# ZEBRA FINCH

Pairs dance while courting, wiping bills and twisting tails, and the male puffs up his more colorful plumage. Then the couple mates for life.

These are social birds, normally found in flocks, but each couple has its own perch on a thorny bush on which to build and maintain a nest of grass, soft stalks, and feathers. The female lays three to seven bluish white eggs, which both parents incubate for about two weeks. Both parents also feed the young until they are ready to leave the nest. Even then, the mother and father lead the chicks back to the nest at night, until they reach maturity when they are two and a half to three months old.

*Taeniopyga guttata*

male       female

ZEBRA FINCH

*Apis mellifera*

# HONEY BEE

THE QUEEN is shown surrounded by her retinue of attendant drone bees, who close in to caress, clean, and feed her whenever she stops to rest.

Mating between the hive's queen and any number of drones takes place in flight, and culminates with the queen ripping out her momentary partner's genital organs, killing the drone in the process. She then returns to the hive to lay her eggs.

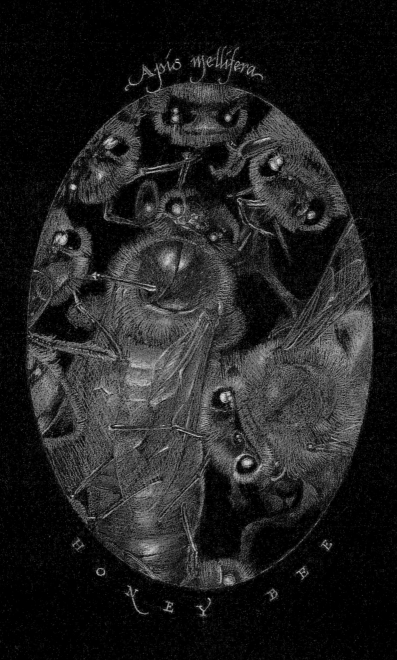

*Apis mellifera*

HONEY BEE

*Equus quagga boehmi*

# GRANT'S ZEBRA

To EARN the right to rule a zebra harem of several mares, and their foals, a stallion must fight and win ritualized battles. Although these fights are full of sound and fury, bites and kicks, injuries are seldom serious.

The winner of a battle will mate with the entire harem until he can no longer defend his conquest. The loser joins a herd of bachelors or tries to abduct a mare from another stallion's herd in an attempt to start his own. Despite battles for supremacy and occasional abductions, harems, or herds, stay together in family units for life.

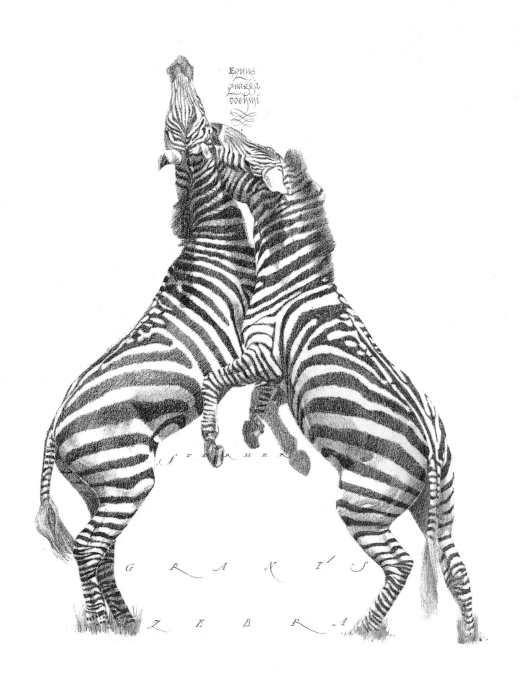

Equus
quagga
boehmi

STORMER

GRANT'S

ZEBRA

*Loddigesia mirabilis*

# MARVELOUS HUMMINGBIRD

Aᴌᴛʜᴏᴜɢʜ ꜰᴇᴡ outsiders have seen, much less photographed, this gorgeous native of the Peruvian Andes, it is truly well named. It has an iridescent gorget, or bib, and, unique to this species, two of the male's four tail feathers grow to four or five times the length of the bird's body.

With these wonderful flags, the male courts the female with a semaphore signal, first raising one feather high over his head, then the other, while hovering directly in front of her. After the female has been seduced, the pair share parental duties, with the male defending the nest. Two eggs are hatched after about fifteen days of incubation, and the chicks become fledglings in about three weeks.

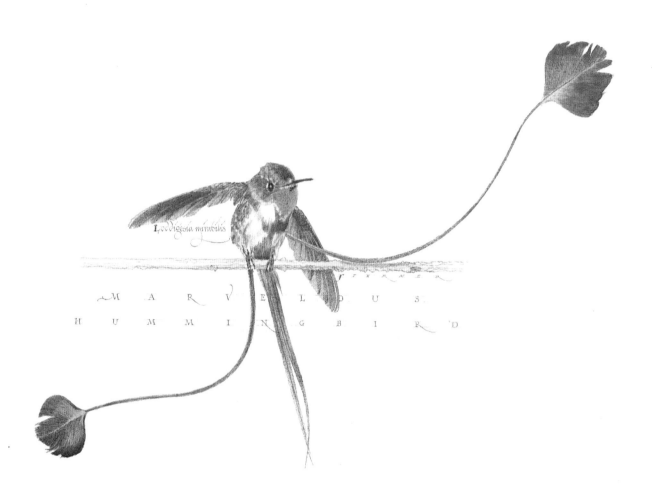

*Loddigesia mirabilis*

MARVELOUS
HUMMINGBIRD

*Homo sapiens*

# WODAABE

INADVERTENTLY creating a
mirror image of an American beauty pageant, the
Wodaabe people of Niger practice an annual ritual in
which the men apply great quantities of facial makeup
and don lavish costume, then prance and dance, teeth
and eyes flashing, in front of the assembled single women,
competing for their attention in the hopes of being
chosen as a mate.

In the marvelous book, *Nomads of the Niger*, by Carol
Beckwith and Marion Van Offelen (Abrams, 1983), the
authors explain, "Among the Wodaabe, beauty is of para-
mount importance, especially for the young men. Because
it is the man's role to attract women, men spend more
time than women adorning themselves and applying
makeup. If a Bodaado [one of the Wodaabe] is not hand-
some, he can compensate for that shortcoming with *togu*.
The word denotes a combination of charm and seduc-
tiveness that has to do with one's way of being with others, one's manner of speaking, the sound of one's voice,
one's sociability, enthusiasm and sense of humor. A man
or woman who has *togu* will attract the goodwill and
friendship of others, and will never be alone. The
Wodaabe say that *togu* comes from the heart: 'He who
has *togu* speaks with his heart.'"

*Homo sapiens*

WODAABE

*Dispholidus typus*

# BOOMSLANG

THE COMMON name means
"tree snake" in Afrikaans. The male is attracted to the
female by her scent. Like most snakes, the male slides
over and around the female, finally orienting himself to
her head-to-head and tail-to-tail to facilitate copulation.

Despite the boomslang's relatively nonaggressive nature,
its bite is poisonous and potentially fatal to man.

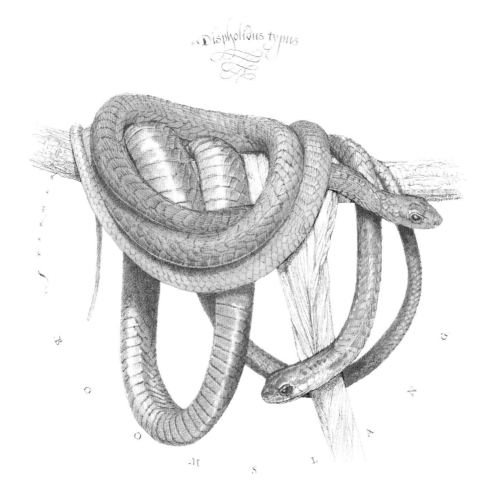

*Dispholidus typus*

B O O M S L A N G

*Ploceus subaureus*

# GOLDEN WEAVER

Attractiveness in this species has much to do with architectural skills and speed. To lure a female, the male weaver must first build a tightly woven, fully domed and protected nest with dispatch as well as dexterity. It must be completed within a week so that the leaves, twigs, grasses, and reeds with which it is built will remain green. If the nest turns brown or yellow, or is found to be otherwise unacceptable to the female, the male must begin again elsewhere.

If the nest is correctly built, it will keep out wind and rain, and be closed tightly against any predator. Three to six eggs will be laid within this nest, to be incubated by the mother for about two weeks. The young are then fed by both parents until they fledge in another fifteen to twenty days.

Ploceus Subaureus

GOLDEN
WEAVER

*Lucanus cervus*

# STAG BEETLE

THE STAG BEETLE is named for the male's extraordinary mandibles, which resemble the antlers of a stag. Like a stag, the beetle uses its "antlers" to defend itself and to engage in battles that establish its right to mate with several females of the species.

This species is the largest beetle in Great Britain; the male grows to a fearsome three inches in length. At one time the head and jaws were carried as a talisman.

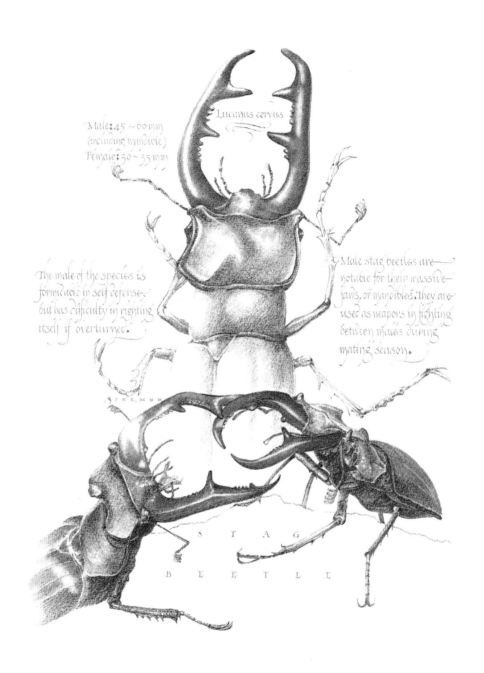

Lucanus cervus

Male: 45 ~ 60 mm
(including mandible)
Female: 30 ~ 35 mm

The male of the species is
formidable in self defense—
but has difficulty in righting
itself if overturned.

Male stag beetles are
notable for their massive
jaws, or mandibles. They are
used as weapons in fighting
between males during
mating season.

STAG

BEETLE

*Gallus gallus*

# RED JUNGLE FOWL

This is the ancestor of all domestic breeds of fowl, and closely resembles the common barnyard rooster. Its original home is six thousand feet up in the forests of the Himalayan foothills.

During courtship the male fluffs up his feathers, lowers his head, and runs in circles around the chosen female with his near wing trailing; then he turns and repeats this ritual dance in the other direction.

G. *gallus* breeds from March to May. Before laying her eggs, the female scrapes together a nest of leaves, usually under a bush or bamboo. She incubates her five or six brownish eggs for about three weeks. Interestingly, while the male is successively polygamous, he will often guard the brood of his most recent mate until the chicks are able to fend for themselves.

*Gallus gallus*

RED JUNGLEFOWL

*Triturus alpestris*

# ALPINE NEWT

This animal is known for its exuberant courting habits and its change of appearance during breeding season. The male's tail fins and dorsal crest become enlarged, frilled, and brightly colored. Displaying all this temporary finery, he prances about the female with extreme energy. When he approaches her, he follows her around with his nose against her side. If she persists in moving ahead of him, he gets around in front and once again performs all of his colorful moves.

After she accepts his advances, his tail vibrates, shaking the frilly crest, and he deposits his spermatophores in a little cluster at the bottom of a pond. She then takes up the clumps into her cloaca, and fertilization is accomplished.

*Triturus alpestris*

STEKMER

female          male

A  L  P  I  N  E  ·  N  E  W  T

*Ectophylla alba*

# TENT BAT

ALSO KNOWN as tent-making bats, these furry ping-pong balls with wings weigh about one-sixth of an ounce and are among the smallest of mammals. They get their common name from the rainproof tents they create with heliconia leaves. White is an unusual color in the bat family, but it renders the little animal nearly invisible through the green of its leafy nest.

The tent bat is shown on the frontispiece as a mating pair. Here the male tent bat rests surrounded by a harem of about twenty females.

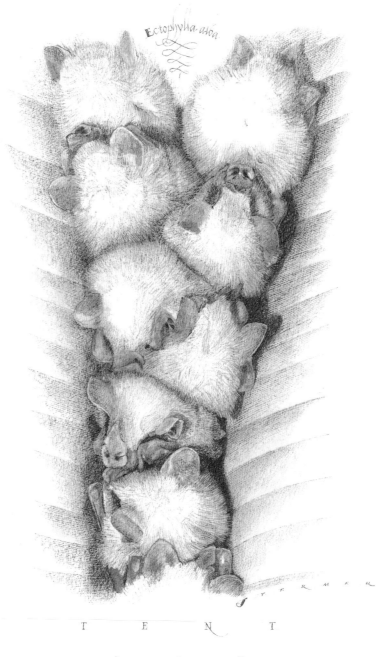

*Ectophylla alba*

STERMER

T E N T

B A T

*Fregata magnificens*

# MAGNIFICENT FRIGATE-BIRD

FRIGATE-BIRDS are nonswimming sea birds capable of extraordinary flights ranging over thousands of miles. During the breeding season, the male selects a suitable nesting site in a tree, on a fallen branch, or occasionally on the ground. He then advertises for a mate by inflating his vivid crimson, balloonlike throat sac, hoping to get the attention of a strong female flying above.

Uttering a falsetto warble, the male spreads his wings throughout the display, his body quivering, his bill and wing quills rattling. If he succeeds in seducing a female, together they will construct a flimsy nest. After the single white fertilized egg is laid, the male hovers nearby during the fifty-day incubation period.

*Fregata magnificens*

female

male

STARMER

M A G N I F I C E N T
F R I G A T E B I R D

*Panthera leo*

# AFRICAN LION

Lions are the most social of all the large cats and live in affectionate family groups called prides. These consist of as few as three to four or as many as thirty individuals. The "king's" position is accomplished and maintained by physical superiority. Having established his status, it is his responsibility to

[*continued on page 38*]

*Panthera leo*

L   I   O   N

# MATING

*Panthera leo*

# AFRICAN LION

[*continued from page 34*]
defend the territory, as well as his position, against any
and all pretenders to the throne. Roaring thunderously
to delineate the pride's boundaries, he then mates with
the lionesses. Both sexes nuzzle, rub, and lick each
other's head and neck. Humming and purring express
contentment or appeasement. Ultimately, males miaow
while copulating.

Though young males must leave the pride when they are
three and a half years old, lionesses—the hunters—often
remain with the same pride their entire lives, providing
food and rearing the cubs.

Panthera leo

L I O N

*Ptilonorhynchus violaceus*

# SATIN BOWERBIRD

THIS IS the only member of the genus, and the entire population is confined to eastern Australia. In lieu of colorful plumage or daredevil aerial displays, the male bowerbird attracts the attention of the female by building an extravagant love nest. This is not used for the laying of eggs or incubation, but only during courtship and mating.

The whole structure—a floor mat of small sticks walled in on all sides—is painted with a concoction of vegetable juices and saliva and decorated with shells, flowers, leaves, and dead insects. After the male bowerbird has exhausted his creativity, he further courts his chosen mate by dancing around with berries in his beak. The combination of the bower, berries, and the dance finally convince the female to mate, which occurs within the structure. Subsequently she leaves her bridal bower to build her plain, ordinary nest elsewhere, there to lay her two eggs and raise her young alone.

*Ptilonorhynchus violaceus*

SATIN
BOWERBIRD

*Hypoplectrus aberrans*

# YELLOW-BELLIED HAMLET

RECENT ADVANCES in audio technology have allowed scientists to understand the behavior of fish to a degree unthinkable only a few years ago. In frequencies far out of the range of human's hearing, fish communicate with each other aurally, warning of danger or inviting sexual encounters. It is thought that the character of the male yellow-bellied hamlet is advertised by means of this voice, and the female in turn communicates to him with a squeal that she is about to release her eggs, signaling that he should prepare to eject his sperm to fertilize them.

This species' common name refers to its color, but the name might also indicate its behavior. The yellow-bellied hamlet hides in the mixed schools of other fish to find protection against predators and to use them as a blind from which to launch surprise attacks on small prey.

*Hypoplectrus aberrans*

YELLOW-BELLIED HAMLET

*Agrion virgo*

# DAMSELFLY

A.K.A. DEMOISELLE AGRION

SOME DAMSELFLIES mate in flight, while others cling to reeds or branches during copulation, which can take only a few minutes or an hour or more. The female may mate with several partners, so the male of the species has developed a scooper with which he can divest her of the sperm of other males. Often, while the female lays her eggs, the male remains attached to her by his terminal appendages, as he does during mating, to ensure that he is the sole inseminator. He may let go of her as she descends to the water's surface and releases her eggs, then clasp her again as she rises, only to repeat the operation somewhere else.

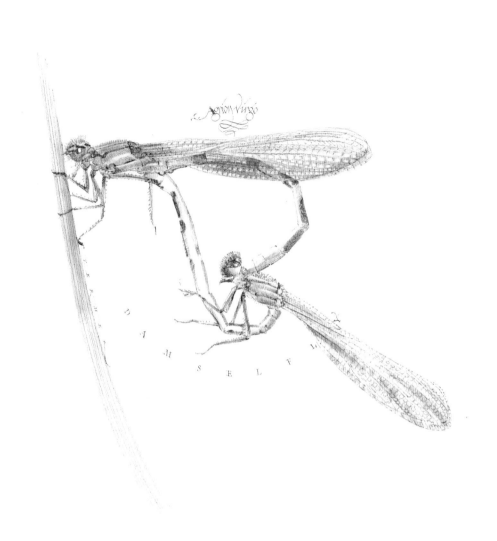

*Agrion Virgo*

DAMSELFLY

*Arctocephalus pusillus*

# SOUTH AFRICAN FUR SEAL

Fᴜʟʟʏ ɢʀᴏᴡɴ ᴍᴀʟᴇꜱ, up
to three times the size of the females, fight among themselves for the best sites at which to mate with their harems.
Younger and older males, unable to defend their territory,
are driven off and play no part in the mating cycle, which
occurs from early November through February.

*Arctocephalus pusillus*

STERMER

South African

F U R   S E A L

*Ariolimax dolichophallus*

# BANANA SLUG

THIS VARIETY of slug carries the Latin surname *dolichophallus*, meaning, roughly, "long penis." Banana slugs are hermaphrodites; therefore, they are equipped with the appropriate male appendage as well as the female organs.

One banana slug begins the thirty-six-hour mating ritual by following the slime trail of another. When the chaser catches up with the chased, the two alternately follow each other, touching and feeling with their radulae, apparently stimulating one another into further, more intense, courtship. Finally, hours later, one takes the dominant role and eats away the mucus plug located in the tail of the other, then they change roles.

Usually by evening, copulation begins and continues through the night and the next day and night, finally ending on the morning of the third day. The denouement occurs when the "female" eats the nearly one-inch-long penis of her mate, sometimes with his help.

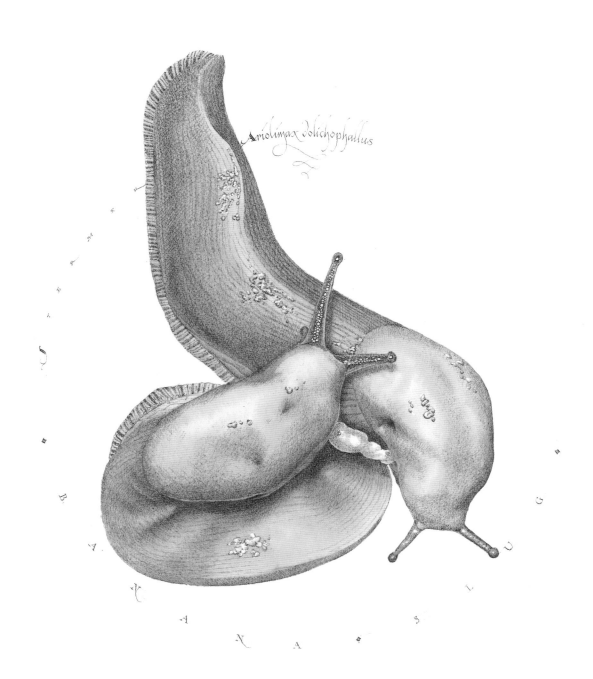

*Ariolimax dolichophallus*

*Pongo pygmaeus*

# ORANGUTAN

THIS IS ONE of the very few species, other than Homo *sapiens,* whose members occasionally mate face-to-face, as shown. After the female finds the largest and strongest male in her area, the couple mates for about a week, during which time the pair will eat and sleep near each other. Afterward, however, the bond is over. The female orangutan will have only three or four offspring in her lifetime, which may last as long as forty years.

*Pongo pygmaeus*

ORANGUTAN

*Edriolychnus schmidti*

# DEEP-SEA ANGLER FISH

THERE ARE three fish depicted here, a female and two males. Perhaps because of the total blackness at the depth in which it lives, the deep-sea angler fish has evolved a unique survival strategy. While still young hatchlings, one or more males attach themselves to a single female by sinking their jaws into her skin, never again to part.

As time goes by, the female's skin grows and envelopes the males, completely absorbing them into her entire life support and reproductive systems. They spend their lives as little more than built-in combination scrotum/penises for the free-swimming female.

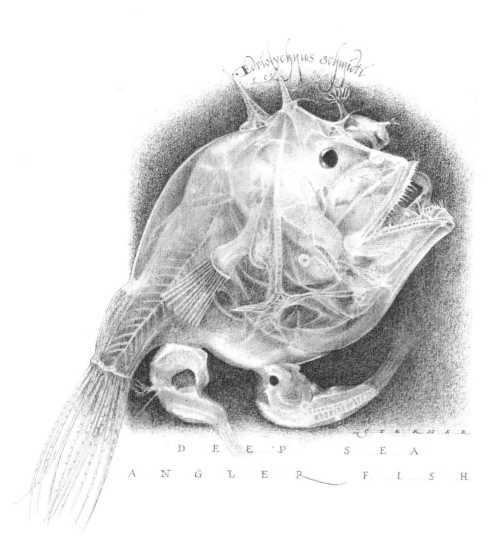

*Edriolychnus schmidti*

STERNER

D E E P   S E A

A N G L E R   F I S H

*Varanus komodoensis*

# KOMODO DRAGON

THIS SO-CALLED DRAGON is, in fact, the largest living lizard. It grows to over ten feet long and to a weight of three hundred pounds, feeding on young deer and wild pigs that they are able to swallow whole, or in large chunks. *V. komodoensis* now lives only on two Indonesian islands, Komodo and Flores.

The Komodo has, in addition to a forked tongue, a forked penis. When a mate is found, the male will mount her with his legs on either side of her body, tuck his tail under hers, and insert one of his hemipenises. Soon after ejaculation he quickly moves on.

*Varanus komodoensis*

KOMODO DRAGON

*Crepidula fornicata*

# SLIPPER LIMPET

THE WONDERFULLY NAMED C. *fornicata* is not a true limpet but a limpet-shaped relative of the periwinkle. They live on top of one another in piles. All slipper limpets start their lives as males, but when one settles on a rock, it becomes a female. Soon a second limpet settles on top of the first, remains a male, and fertilizes the first. As each new individual comes to live on the pile, the male it lands on begins to change its sex to female, after which it is fertilized by the male newcomer. Thus, the pile—or chain—of slipper limpets is always composed of many females below, at least one male on top, and those limpets between in the process of changing from male to female.

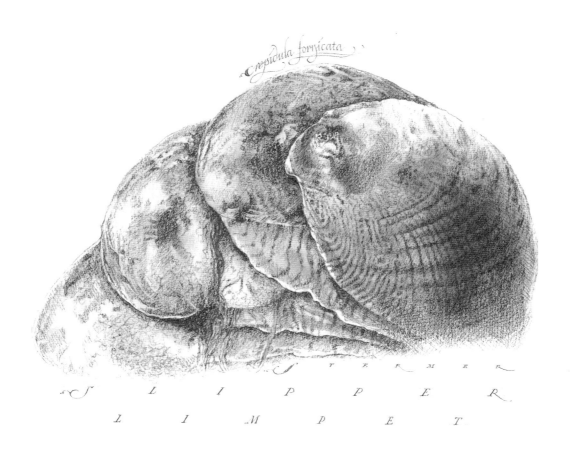

*Crepidula fornicata*

S T E R M E R

S L I P P E R

L I M P E T

*Diceros bicornis*

# BLACK RHINOCEROS

Courtship among black rhinos is nothing nice. The wonder is that the species continues to reproduce. First the female in estrus and the attending male circle warily around each other, she whistling shrilly, he sighing, until the male turns and charges the female; since they both weigh in the neighborhood of three thousand pounds, she feels no need to back off. They continue to run head-on into one another, butting and hooking their enormous horns. This violent foreplay has been known to prove fatal; zoos are understandably reluctant to breed rhinos, not wishing to risk losing one in the process of gaining another.

As weariness takes over, the female may decide that the male is worthy to sire her offspring, and after a few more parries she offers herself to the male, rear first. He mounts her clumsily (imagine copulating Buicks) and attempts to penetrate her with a penis measuring about two feet long. He may remain mounted, climbing ever higher and penetrating ever deeper, for as much as an hour, ejaculating five or six times during that period. When he finally withdraws, both animals are completely exhausted, but after a short period of rest, the whole process is repeated.

The female will be pregnant for about fifteen months, after which she will heave, sway, swell, and bellow until her 130-pound calf is all but ejected out of her body.

*Diceros bicornis*

B L A C K

R H I N O C E R O S

*Chelonia agassizi*

# BLACK TURTLE

COPULATION can last for hours, with the male gripping tightly to the female, even during deep dives. While the male black turtle lives out his entire life at sea, the female comes ashore to lay her clutch of up to eighty-five eggs.

This species may be a subspecies of the green turtle and is found only in the southeastern Pacific Ocean, ranging from Baja California to the Galápagos Islands.

*Chelonia agassizi*

BLACK · TURTLE

*Enhydra lutris*

# SEA OTTER

VIOLENCE during mating is not unusual in the animal kingdom; witness the cannibalistic tendencies of the praying mantis and the black widow spider. But less well known, and more surprising, is the pain inflicted by the male sea otter upon his mate.

To hold her still while mating in water, the male grasps her with his claws and bites her on the nose as she lies on him with her head bent back. The bites can be serious and nearly always leave permanent pink scars. This disfigurement helps field scientists identify mature females.

*Enhydra lutris*

SEA OTTER

*Everes comyntas*

# EASTERN TAILED BLUE

UNLIKE MANY of its relatives, this species of butterfly is highly adaptable to changing conditions, and even increases its population amidst the expanding activities of man, along roadsides and in city parks and yards. Many of the plants that provide a home for the butterfly are cultivated varieties.

The life cycle of E. *comyntas* begins with copulation, during which the male deposits his sperm into the female while the two are held together with claspers at the end of the male's abdomen. The act can last from thirty minutes to several hours. Not long afterward the female lays several hundred eggs upon a carefully chosen flower. In a little more than a week, caterpillars emerge, beginning their renowned metamorphoses into butterflies.

*Everes comyntas*

EASTERN TAILED BLUE

*Felis concolor*

# MOUNTAIN LION

F. CONCOLOR is known by many common names, depending on the species' geographical location. Puma is the word used in Mexico and South America; cougar, catamount, and mountain lion are the most common names in the United States and Canada.

Females can come into estrus during any season, but will usually give birth only once every two years. The gestation period for mountain lions lasts from ninety to ninety-eight days. Cubs stay with their mother for a year to a year and a half, during which time she protects them, sometimes even from the father, and trains them to hunt. Eventually they leave to stake out their own territory.

*g Felis concolor*

M O U N T A I N • L I O N

*a.k.a.*

C O U G A R

*a.k.a.*

P U M A

*Giraffa cameloparialis*

# GIRAFFE

GIRAFFES ARE DOCILE and peace-loving. They traditionally live in herds of up to twenty or thirty members, in which females far outnumber males. Old males are exiled and live in isolation.

During the breeding season, the dominant bull—so decided by nonviolent ritual combat—mounts the receptive cow of his choice and then returns to the other males in the herd, having nothing further to do with the pregnant cow or her offspring.

After a gestation period of about fourteen months, a single calf is born, six feet tall, weighing about 130 pounds, and able to walk within an hour or so after birth.

*Giraffa camelopardalis*

STERMER

G I R A F F E

*Tetraopes tetraophthalmus*

# RED MILKWEED BEETLE

THE MALES AND FEMALES of this species usually meet each other by accident. Occasionally an attractive chemical scent is released by the female, but this seems to be an optional procedure. When members of a pair find each other, the male gropes at his intended with his antennae, often hitting her soundly, testing her for interest and possibly for species identification. The two may be observed gnawing at each other with their powerful jaws to further confirm identity.

Once this odd courting has been completed, the female will bore a hole in nearby bark, to act as a repository for her eggs, and copulation will occur.

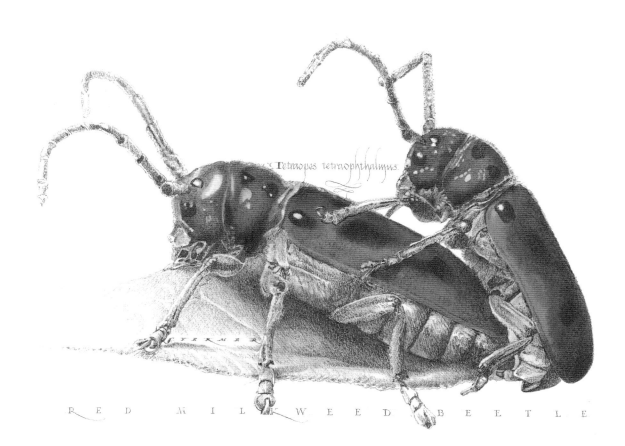

Tetraopes tetraophthalmus

RED MILKWEED BEETLE

*Pandion haliaetus*

# OSPREY

THE OSPREY is normally monogamous, at least during the breeding season. It is certainly faithful to its breeding location, returning to it year after year, and often returning to its mate, as well.

Courting requires several weeks of sky dancing. The male repeatedly dives and then stops, occasionally flying backward, all in an apparent attempt to mark his territory, to repel intruders, and to impress his intended. As the sky dancing abates, the male will bring fish as an offering during a ritual feeding.

After courtship is completed, copulation occurs. The female lowers her wings and fans out her tail; the male lets out shrill calls and then lands on her back. The female answers his song with one of her own, creating an avian duet.

Osprey first breed when they are three years old, can live for about twenty years, and copulate annually. The male's responsibility doesn't end with mating, however. He must bring food, mainly fish, to his mate while she incubates the eggs, and they both feed the young until they fledge.

*Pandion haliaetus*

OSPREY

*Cyerce  nigricans*

# FRILLED  SEA  SLUG

These brightly colored underwater animals, also known as nudibranchs, are true hermaphrodites. They mate reciprocally, exchanging both sperm and eggs, following which each partner becomes pregnant. Some sea slugs lay fertilized eggs just ten days after they themselves have hatched.

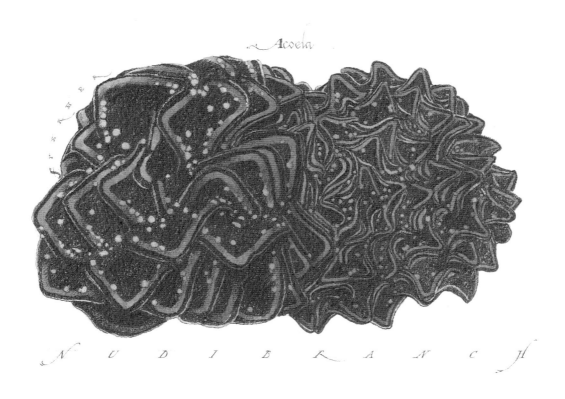

Acoela

N U D I B R A N C H

*Ursus arctos*

# BROWN BEAR

WHEN THE brown bear is busy with its solitary activities—sleeping, eating, hunting, and bathing—it joins other bears only for the very short mating season between May and July. The male never hangs around during the gestation period of 180 to 250 days, nor does he take any part in raising the cubs.

In December or January, during hibernation, birth takes place. While the cubs suckle and the mother cares for them, she neither eats nor drinks. At the end of hibernation, the mother and cubs emerge hungry, and the newborns begin their training.

Ursus arctos

B R O W N   B E A R

*Limax maximus*

# GREAT GRAY SLUG

These invertebrate hermaphrodites circle and rub against each other on a branch for up to an hour and a half, exchanging mutual caresses with their tentacles. During this foreplay, they secrete mucus that forms into a cord. The slugs then hang from this cord, entwining and twisting, all the while extruding from their heads blue penis sacs that end in fan-shaped masses. After the transfer of sperm is accomplished, the slugs climb back up the cord, which is then eaten, and they slide along the branch and down the tree to their routine earthbound life.

*Limax maximus*

GREAT SLUG

Sterner

*Tenodera  aridifolia  sinensis*

# CHINESE  MANTIS

CLEARLY LOVE, or even mutual affection, often has little or nothing to do with the propagation of any species. This is especially true in regards to this species. If the male is not extremely agile and quick, he is likely to provide lunch for the larger, more voracious, female.

The male must wait until he thinks his mate cannot see him (tough, considering that mantids have exceptionally good eyesight) then he pounces on her back, fits his forelegs into grooves at the base of her wings, and grips. His long abdomen curves down until his penis finds her reproductive opening, wherein he pumps sperm into her body.

A few lucky males quickly escape to safety. The others, however, are decapitated by the female's strong jaws. While she munches away, he continues copulating, until he finishes or she finishes him, whichever comes first.

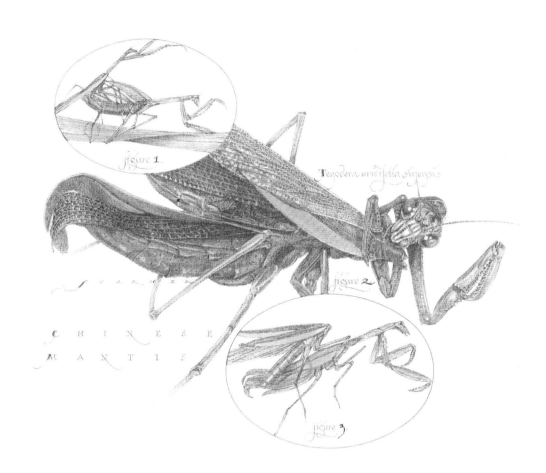

figure 1.

Tenodera aridifolia sinensis

figure 2.

S T A R M E R

C H I N E S E

M A N T I S

figure 3.

*Macropus giganteus*

# GRAY KANGAROO

THE FIRST field biologists to study the kangaroo, accustomed as they were to mammalian life on other continents, must have been astonished and even bewildered by the marsupial's reproduction system. The female has three vaginal tracts, while the male has a penis that lies behind his testicles and points backward when erect. The embryonic offspring leaves the uterus about a month after conception and, when just about the size of a kidney bean, attaches itself to a nipple inside its mother's pouch, where it continues to develop for 190 days. It will have grown several thousand times its birth size before it leaves the pouch, and even then it will return—and be welcomed back—for months thereafter.

One point of behavior among kangaroos is similar to that of many other mammals: The boomers (males) will fight for dominance within a mob (social grouping) and the attendant right to mate with all the fliers (females). To insure some genetic diversity, however, other males may sneak a copulation while the dominant male is occupied elsewhere.

*Macropus giganteus*

GREY KANGAROO

*Octopus vulgaris*

# COMMON OCTOPUS

"FRAUGHT WITH PERIL" describes the mating of octopi. If the male is not patient and properly deferential, the female may attack him, dismember him, and even kill him. Maybe she feels entitled; the act of copulation will be her last, and after her eggs are laid she dies.

When the male first spots a female with whom he intends to mate, his normally gray body becomes striped and two knoblike horns grow from the top of his head. He first presents himself to her, underside up, then reaches his reproductive arm (acting as a penis) into her breathing and propulsion funnel. For several hours they remain enjoined, while his sperm, enclosed in packets, slip from under his head, through a groove in his arm, and into the mantle cavity of the female, where seawater ruptures the packets and the sperm mixes with her eggs.

The fertilized eggs are then washed from her body and attach themselves to rocks under the water until they hatch. During this period—up to nine weeks—the mother eats very little in her zeal to guard the eggs, and when the offspring emerge, she dies.

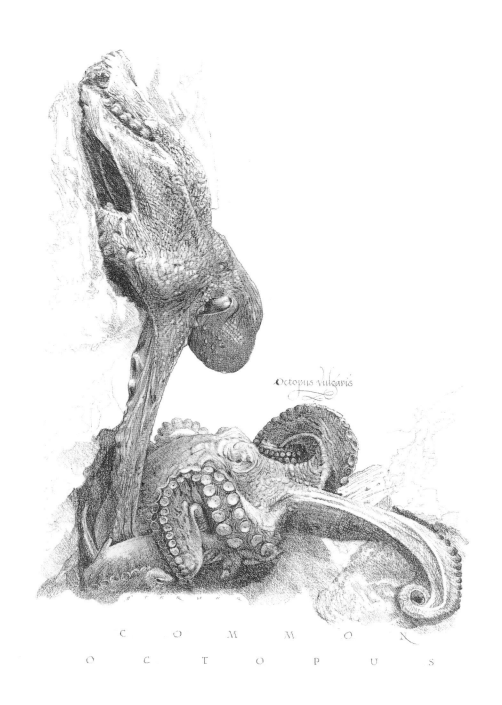

*Octopus vulgaris*

C O M M O N

O C T O P U S

*Pan paniscus*

# BONOBO CHIMPANZEE

I<small>T IS PROBABLY</small> an oversim-
plification to say that P. *paniscus* lives for sex, but if any
species on earth fits that description, this one does. Sex is
not only the means for carrying on the line, it permeates,
if not occupies, every waking moment. Every member of
the group—young, old, male, female—is a willing sexual
partner for every other member almost all the time. Chil-
dren, like those pictured opposite, play at copulation long
before they are mature enough to have experienced its
greater implications. And, most remarkable of all, sex
seems to have entirely replaced violence in this species'
behavior patterns. Bonobos, or pygmy chimps, don't fight,
either among themselves or with others, nor do they kill
prey, except as an aberration.

These natives of Zaire are rarely observed, and only
recently have their features and behavior been studied
sufficiently to cause scientists to name them a distinct

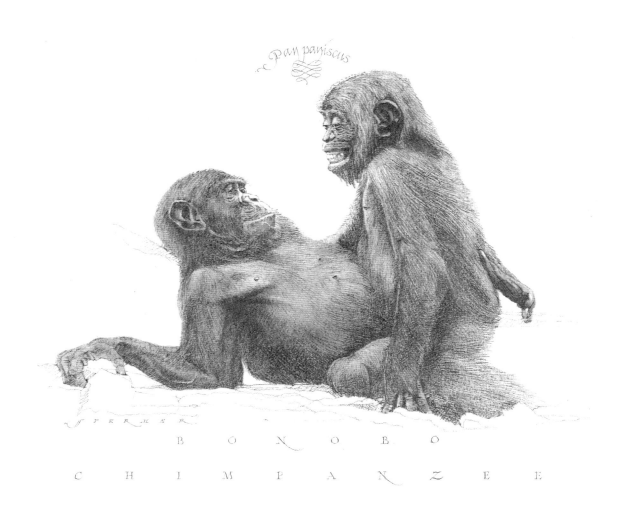

Pan paniscus

BONOBO

CHIMPANZEE

species. Bonobos are considered an older species than the other species of chimps, and may resemble the common ancestor of humans and apes. Does that mean that our violent ways are the product of evolution and that our evolutionary ancestors were far more humane than Homo *sapiens* is now?

Although females may be lower in the social hierarchy than males, they are the initiators of sexual behavior, often by offering their rear ends to the males, and they control the positioning. Males often offer their genitals to other males and rub their buttocks together as a sign of peaceful intent. As further proof of their absorption with sex, bonobos have at least twenty gestures and calls expressing the willingness to copulate.

*Pan paniscus*

STERNER

B O N O B O

C H I M P A N Z E E

*Gygis  alba*

# FAIRY  TERN

THIS LOVELY little member of the tern family wears pure white, except for a ring of black around the eyes and another around the legs. It is distinguished by a charming courtship ritual in which the male presents a fish to a prospective female. It is believed that the female selects the better fishers as mates, insuring that her offspring will inherit the essential genes that determine this skill. Mating with a successful fisher is important because the female stays by the nest prior to laying her eggs, and the male brings her food during the gestation period.

G. *alba* has a curious, if not unique, nesting habit. The female lays her one egg balanced precariously on the bare branch of a mangrove tree or low bush. The parents alternate incubation duties with extreme caution, lest their only offspring become dislodged unto destruction. After hatching, the chick hangs onto the branch, sometimes upside-down, for several days before falling or jumping to the ground.

Gygis alba

Sterna

FAIRY TERN

*Paramecium*

# PARAMECIUM

THIS MICROORGANISM inhabits any and all kinds of water, including ponds, streams, lakes, tide pools, oceans, and sulphur springs, as well as the water found in our bodies. They have two kinds of nuclei, the macronucleus and the micronucleus. Although they are single-celled creatures, they are quite complex.

This species can reproduce both sexually and asexually. Male and female terminology is irrelevant because paramecia are identical in structure. When two join, they exchange portions of their micronuclei in a process called conjugation. Then they separate and continue to reproduce by cell division. If more nuclei are needed, the cell undergoes meiosis, and the nucleus splits to become two nuclei.

Should a paramecium find itself without an available partner, the extra nucleus will fuse with the other nucleus, called autogamy. The cell will then split and meiosis will occur.

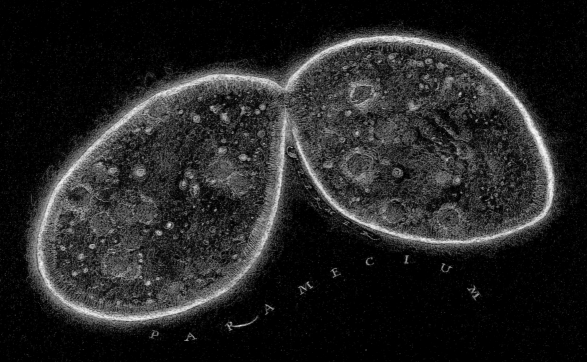

PARAMECIUM

*Phyllomedusa trinitatis*

# CARIBBEAN FROG

THE MATING INSTINCT of these arboreal amphibians reaches its steamy climax during the rainy season in the American tropics. They don't hop. They crawl to, and then climb, low bushes beside pools, where they mate among the leaves overhanging the water. The female shapes a leaf into a tube, and then lays her eggs inside as the male fertilizes them. Occasionally this becomes so frenzied that group sex, known as multiple amplexus, occurs.

Hatching is also a highly charged event. High pressure builds up within the membrane, resulting in little explosions that hurl the tadpoles into the water below.

*Phyllomedusa tristatis*

CARIBBEAN · FROGS

*Loxodonta africana*

# AFRICAN ELEPHANT

PREPUBESCENT ELEPHANT calves play at sexual activity at a remarkably young age. However, adults seem to experience far less pleasure than the cavorting youngsters. The bull elephant usually has to fight and defeat another bull before he earns the right to chase a reluctant cow and finally mate with her.

The actual act takes less than a minute; gestation lasts twenty-two months.

*Loxodonta africana*

E A L F E R P I H C A A N N T

*Presbytis  entellus*

# L A N G U R

P. ENTELLUS IS the sacred langur of the Hindus, and unlike its cousins in the genus, it is a largely terrestrial monkey. Langurs are a comparatively peaceful species; changes in dominance from one male to another occur without violence, and infidelity is looked upon with indifference. Even members of established territories accept visits from outsiders.

The langurs live in what we might consider well-adjusted family groups. The birth of a single young monkey occurs after a gestation period of about six and a half months. The young are watched over by the mother until she becomes pregnant again some time late in the following year.

*Presbytis entellus*

FIGURE 1

L A N G U R

FIGURE 2

*Schistocerca gregaria*

# DESERT LOCUST

Asingle swarm of desert locusts can occupy a space as large as two thousand square miles and travel more than three thousand miles in less than two months. It is, therefore, difficult to speak of them in groups as small as mating pairs, or, in fact, to think of them reproducing at all. But reproduce they do, with astonishing efficiency.

A swarm ceases its migrations when environmental conditions are favorable to mating; courtship then begins. The male approaches a receptive female, feeling at her with his antennae. Then, with wings folded, he pushes from behind and grabs her with hooks on his genitals. He ejects his sperm into her reservoir, and the act is completed, lasting just under an hour.

The female can lay about five pods in the soil in a lifetime, each pod containing up to 120 eggs. The insects that emerge are embryonic in appearance and are called pronymphs. They subsequently go through six growth cycles—or instars—before maturity.

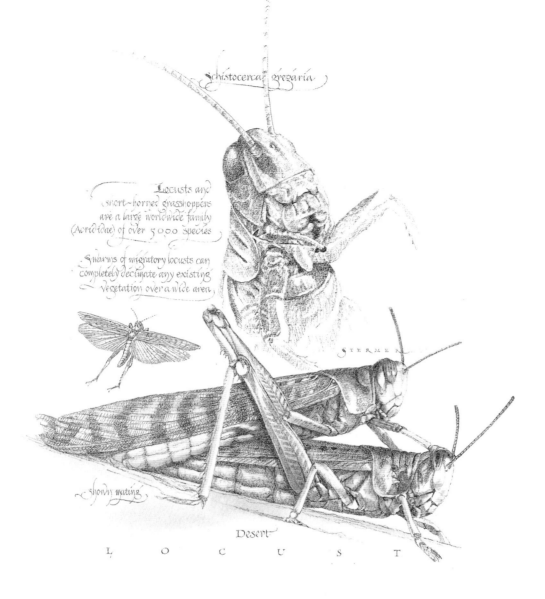

Schistocerca gregaria

Locusts and
short-horned grasshoppers
are a large worldwide family
(Acrididae) of over 5000 species

Swarms of migratory locusts can
completely decimate any existing
vegetation over a wide area.

STERMER

Shown mating

Desert
L  O  C  U  S  T

*Geochelone elephantopus*

# GALÁPAGOS GIANT TORTOISE

MALE GIANT TORTOISES must engage in ritual combat, during which they stand as high as possible with jaws agape. After the confrontation the victor courts the female by nibbling at her toes. As foreplay, the tortoises spend several hours bobbing their heads at one another.

The shell of the giant tortoise presents almost insurmountable problems during mating. In order for the two to actually mate, the female must stick her rear end out of the shell as far as possible. In case she should demur, the male circles around and snaps at her head, which she pulls in defensively. This forces her tail to emerge at the other end, and he can subsequently mount her. Once they have thus joined, they may remain together for as long as a full day, occasionally letting loose with a resounding roar.

*Geochelone elephantopus*

GALAPAGOS GIANT
TORTOISE

*Balanus balanoides*

# ACORN BARNACLE

THE MALE ORGAN of this barnacle is the largest, proportionately, of any living thing's. Since the species is both stationary and hermaphroditic, barnacles must hunt their temporary mate by means of these extraordinary appendages. If, as occasionally happens, a neighbor is either unwilling or unavailable, self-fertilization is possible.

Eggs are released once a year, and the event is advertised by the release of a chemical that stimulates the neighbors. Sperm is produced in great quantity so that one male can fertilize many other barnacles in a short amount of time.

*Balanus balanoides*

ACORN

BARNACLE

*Sula nebouxii*

# BLUE-FOOTED BOOBY

THE REPRODUCTIVE CYCLE
of this bird coincides with the migration of the Pacific
Ocean fish that provide the main part of its diet. The mating ritual begins with a dance; the female rocks from side
to side, lifting each brightly colored foot in turn, while the
male spreads his wings and whistles. Then together they
both build a crude nest on bare, rocky ground.

After mating, they continue to huddle together affection-
ately until the eggs are laid. It is only a matter of a few weeks
between mating and the hatching of eggs, but still the mating pair are bonded enough to stay together to rear the
chicks; one parent will guard them while the other catches,
then regurgitates fish into the hatchlings' beaks.

Sula nebouxii

STERMER

BLUE   FOOTED

B   O   O   B   Y

*Chiromantis xerampelina*

# GRAY TREEFROG

UP TO THIRTY of these south-
ern African frogs, both male and female, group together
in trees to beat their eggs and seminal fluid into a foamy
nest with their feet. This froth later hardens to protect
the development of the eggs incorporated therein. Then,
during the larval stage, the nest drops into the water, and
the young frogs complete their infancy.

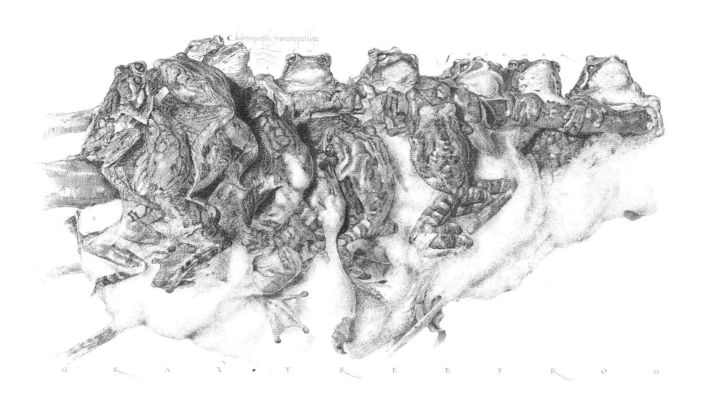

*Chiromantis xerampelina*

G R A Y   T R E E   F R O G

# PROCREATION

*Ara  ararauna*

# BLUE  &  YELLOW  MACAW

Mark Twain wrote, "Man is the only animal that blushes. Or needs to." If the cause is embarrassment, perhaps. However, anxiety can turn members of this parrot family furiously red-faced, as when a pair protects its nest from intruders. They mate for life, both parents feeding and rearing their chicks.

*Ara ararauna*

BLUE AND YELLOW
MACAW

*Hippocampus whitei*

# SEAHORSE

NEARLY EVERY act involving reproduction and rearing in the seahorse's life reverses our conventional assumptions about parental roles. During initial courtship the male and female drag their tails along the floor of the sea, the male with his head tucked to his chest. Soon the female circles the male, displaying her magnificent colors. Finally she grasps him with her prehensile tail, and as they rise to the surface she probes at him with her "penis," actually a cloacal appendix, until she finds his brood pouch. While swimming face-to-face, she ejects up to 600 eggs into the pouch and, with no further responsibility, she leaves the new father to carry, birth, and then nurture their young.

The presence of the eggs in the male's pouch prompts him to release sperm to fertilize them, and a few weeks later he undergoes spasmodic contractions that eventually expel the tiny seahorses into the watery world.

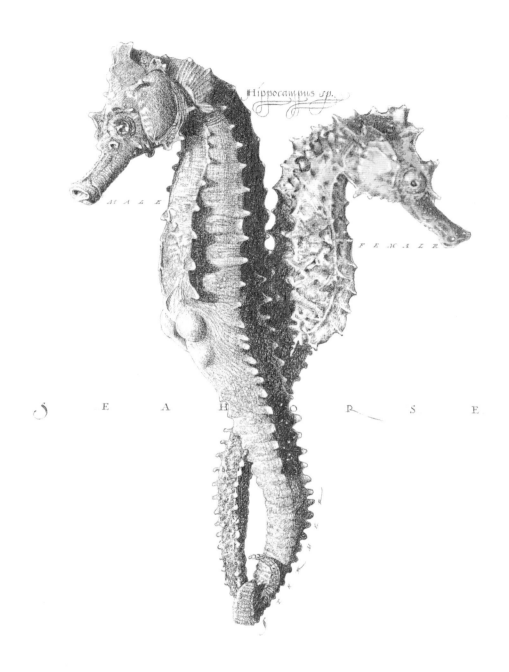

*Hippocampus sp.*

MALE

FEMALE

SEA HORSE

*Haliaeetus leucocephalus*

# BALD EAGLE

Unlike more adroit avian architects—the weaverbird, for example—eagles make little attempt to construct tight nests. They simply pile sticks and branches on a tree fork or rock ledge, relying on their weight to stabilize the mass.

Before undertaking such domesticity, however, courting pairs indulge in some of the most dramatic and perilous activity known in nature, involving spinning aerial descents with talons locked, heart-stopping dives one over the other—sky dancing—all the while calling to each other in high-pitched shrieks.

The female will typically lay two eggs, several days apart. Incubation will last a little over a month, after which the chicks are hatched over a ten-day period. The chicks will then pass through four stages of development at a slow rate of maturation before they reach adulthood. In fact, eagles do not acquire their signature white heads until they reach the second stage of maturation between two to four years of age.

*Haliaeetus leucocephalus*

B A L D  •  E A G L E

*Buthus maroccanus*

# SCORPION

MATING AMONG scorpions begins with a lengthy dance that ends when the male drops the spermatophore then manipulates the female so that her reproductive opening is directly over the mass. Her underside catches and collects the semen; the young are later born fully developed. Regardless of their state of development, young scorpions spend a good part of their infant lives riding on their mother's back before venturing out on their own.

*Buthus wallicranus*

*SCORPION*

*Felis catus*

# ABYSSINIAN [DOMESTIC] CAT

THIS BREED is the closest relative to the cats worshipped by the ancient Egyptians. It often looks latently wild and unpredictable, but is in fact exceedingly affectionate.

Domestic female felines come into heat twice a year, and signal the event by giving off scents that attract males, apparently over a neighborhood-wide area. Mating takes but a few seconds and can occur repeatedly for three to five days. As a result, a female may give birth to a litter of kittens, and each kitten may have a different father.

Kittens take about sixty-five days to gestate. After birth, they will suckle until they are weaned at six to eight weeks of age. Nearly all kittens stay with their mother for some time after they are weaned; in the case of domestic breeds the relationship may last for two years or more.

*Felis catus*

ABYSSINIAN

(DOMESTIC CAT)

*Phoca vitulina*

# HARBOR SEAL

THESE ARE fairly small members of the seal family, seldom weighing over 250 pounds or exceeding five and a half feet in length. The name reflects this animal's tendency never to stray far from land.

The young are born, after a gestation period of nine to twelve months, with white furry coats, which quickly turn spotted. The young nurse for a month, and then are taught to swim and find their own food.

*Phoca vitulina*

Scammon

HARBOR SEAL

( nursing )

*Diomedea epomophora*

# ROYAL ALBATROSS

Monogamy is rare in the animal kingdom. A notable and extraordinary exception is the royal albatross.

A typical relationship may progress as follows. Upon returning from months at sea, adolescent albatrosses—about five years old—gather for a dance near the spot where they were hatched. Eventually males and females pair off and spend the balance of their time in each other's company. After a few weeks they fly off separately and spend a year at sea hunting for food.

They return to the same spot and spend the entire time ashore in each other's company. This ritual is repeated annually until, after four years of courtship, the male picks out a nesting site and the relationship is consummated. Both parents share responsibilities during the seventy- to eighty-day incubation period, one sitting on the egg while the other hunts for food over the ocean. During the chick's first year of life, the parents continue to alternate between raising the chick and searching for food until the offspring is able to fend for itself.

Royal albatrosses live from sixty to eighty years, with a very low mortality rate, about three percent. Pairs are together for a long time, breeding and rearing many offspring during their most unusual and marvelous partnership.

*Diomedea epomophora*

R O Y A L
A L B A T R O S S

*Chamaeleo jacksonii*

# JACKSON'S CHAMELEON

Despite this pose, illustrated opposite, the infant Jackson's chameleon is independent at birth. It is merely resting on one of the horns of the father. The mother, who has a single vestigial horn, can carry up to forty eggs; only ten or so offspring, however, normally survive to birth.

*Chamaeleo jacksonii*

JACKSON'S
CHAMELEON

*Lepus californicus* ssp. *texianus*

# TEXAS JACKRABBIT

THE TERM "March hare" perhaps has its origin in the courting behavior of the male jackrabbit (actually a hare), which begins in early spring each year. The males gather in a circle and perform a dance, during which they move toward the center, leap into the air, scatter, and then return to the circle to repeat the process. They also take part in ritual battle, standing on their hind legs, boxing their opponents, and, when the opportunity presents itself, kicking like kangaroos.

About six weeks after mating, the pregnant female digs an oval nest in a sheltered area, lines it with fur from her own coat, and gives birth to as many as eight offspring. Within a few weeks, during which they are protected by the mother, the young become independent.

Lepus californicus [ ssp. texianus ]

The jack rabbit can jump
up to 5 feet high and
20 feet long.

Infant

Ears:
4¾ inches

Length: up to 2 feet
Weight: up to 12 lbs.

T E X A S
J A C K R A B B I T

*Papio hamadryas*

# HAMADRYAS BABOON

IN THE MOUNTAIN country west of the Sahara Desert, this monkey is known as the Sacred Baboon. It lives in tribes led by the alpha male, an absolute tyrant as long as his strength and agility hold out against all challengers. These tribes number up to a couple of hundred members, so leadership is a perilous occupation.

Mates are jealously guarded. Whether monogamous or polygamous, the male will repel interlopers, even to death. Infidelity is a capital offense, but even after the male inflicts his retaliation, he will guard his dead mate's body as assiduously as if she were alive.

Offspring are strictly reared, primarily by the mother. There is a complicated hierarchy within each troop, and the young must learn their place in the society if they are to survive. Females reach maturity in about four or five years but it takes the males another couple of years to reach adulthood.

*Papio hamadryas*

H A M A D R Y A S
B A B O O N

*Panthera uncia*

# SNOW LEOPARD

This gorgeous cat, also known as the ounce, is not a true leopard. Because it can purr but not roar, and because its feeding habits are similar to those of smaller felines, it actually fits somewhere between the smaller wild cats and the genus *Panthera*.

The snow leopard is exceedingly solitary and elusive, and quite endangered as a species. They were probably never very numerous, and even that small population has been critically depleted by poachers seeking the cat's magnificent coat. Snow leopards pair up only to court and mate; however, pairs may occupy adjacent hunting areas and even occasionally cooperate in ambushes.

Mating happens in late winter, and the two to four cubs are born in April and May. They stay with their mother, first in a den, and then out on hunting forays, through the next winter. They are affectionate with their young, and seem to be conscientious parents, despite their solitary natures. Oddly, unlike many other mammals with more social instincts, the snow leopard adapts to captivity with relative equanimity, even breeding with ease. Nevertheless, its prognosis in the wild is grim.

Head & Body Length: 3¼ ~ 4¼ feet
Tail Length: 2½ ~ 3¼ feet
Shoulder Height: 2 feet
Weight: 55 ~ 165 pounds

SNOW LEOPARD
• Panthera uncia •

*Trichechus manatus*

# WEST INDIAN MANATEE

MANATEES ARE SHY and elusive creatures, but affectionate and even demonstrative among their own kind, especially during courtship. Individual manatees may rise to the surface and nuzzle face-to-face for minutes at a time. While actual copulation is accomplished with considerable splashing about—males weigh about fifteen hundred pounds, females, nine hundred—they are exceedingly peaceful animals. They have virtually no defensive skills and will not fight, either to protect their young or themselves.

Mating occurs in December and the gestation period lasts about thirteen months, after which one or two calves are born, each weighing about sixty pounds. They are born underwater and are able to swim almost immediately. They must learn to breathe properly, however, so for about a week the mother will raise her offspring to the surface every three or four minutes. To nurse, as shown, the calf must hold its breath. Calves are weaned at eighteen months of age, but may still follow the mother around for another six months, finding something else to do when the mother begins her December mating ritual.

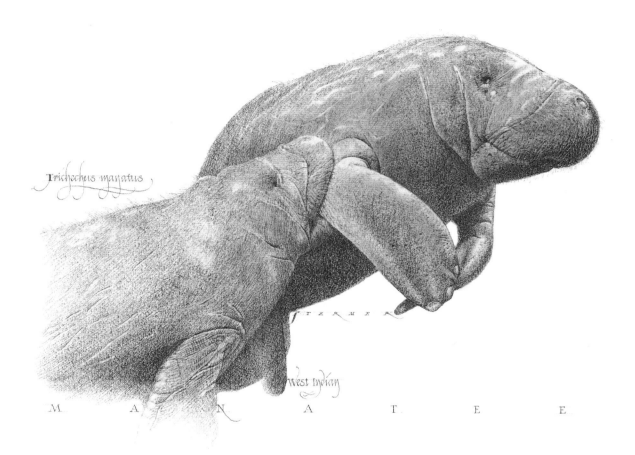

*Trichechus manatus*

West Indian
M A N A T E E

*Tyto  alba*

# BARN  OWL

BECAUSE THE MALE of this species plays such a large role in providing for the young after they are hatched, he first attracts his mate by bringing her rodents. Although clutch size is dependent on the food supply, the female barn owl will lay five to seven eggs over a two- or three-day period and incubate the eggs for about a month. After the young begin feeding, they start to distinguish the smell, sound, and appearance of prey, then mimic their parents and thus learn how to hunt.

· TYTO alba ·

Length: 36 ~ 51 centimeters

Barn owls are extremely
efficient nocturnal hunters,
capable of taking many rodents,
such as mice and rats. They live in
tropical and temperate climates
on all continents.

Lays 5~11 round, white eggs

STERMER

B A R N · O W L

*Archilochus colubris*
## RUBY-THROATED HUMMINGBIRD
AN AVERAGE of two eggs in a nest of plant down, fibers, and bud scales attached to a tree limb with spider silk.

*Dumetella carolinensis*
## GRAY CATBIRD
THREE TO FIVE eggs in a bulky, deeply cupped nest of twigs, grapevines, leaves, grasses, paper, and weed stems that is lined with rootlets.

*Coccyzus erythropthalmus*
## BLACK-BILLED CUCKOO
TWO TO THREE eggs on a platform of small twigs loosely interwoven, lined with catkins, cottony fibers, and dry leaves.

*Gallinula chloropus*
## COMMON GALLINULE
BETWEEN SIX and seventeen eggs in a cuplike nest well constructed of dead cattails, rushes, and stems.

*Pelecanus occidentalis*
## BROWN PELICAN
UP TO THREE eggs in a nest built sometimes on the ground but mostly in trees. Nest made of sticks, reeds, grasses, and other available materials.

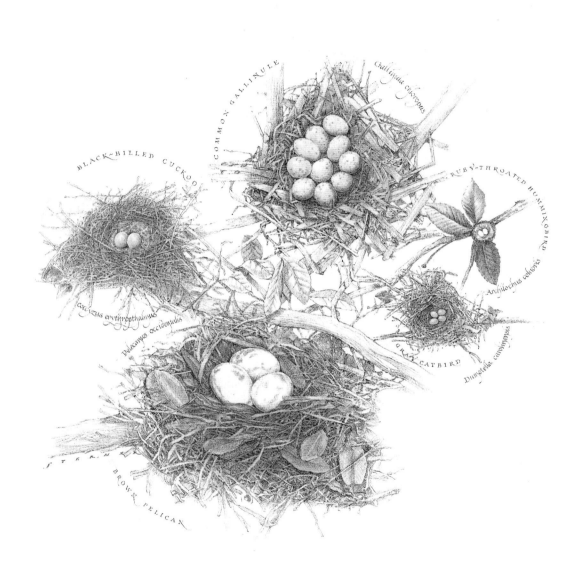

BLACK-BILLED CUCKOO

COMMON GALLINULE

*Gallinula chloropus*

RUBY-THROATED HUMMINGBIRD

*Coccyzus erythropthalmus*

*Archilochus colubris*

*Pelecanus occidentalis*

GRAY CATBIRD

*Dumetella carolinensis*

BROWN PELICAN

# A C K N O W L E D G M E N T S

MANY PEOPLE were help-ful during the course of my work on this book. One was essential throughout. Megan Blue Stermer, my daughter, was indispensible as a researcher, com-piler, editor, cheerleader, and coura-geous critic. Jenny Collins, editor at Collins, kept the book on track with grace and wit, and Tom Morgan, Blue Design, gave it form. Without these three, the book would likely have remained orbiting in my head. And if it weren't for my wife, Jeanie Landles, it would have lacked a title.

I also owe a debt to various individuals, friends really, who gave generously of their expertise, experience, and goodwill: Pieter Folkens; Keith Howell, editor of *Pacific Discovery*; John McCosker, director of the Steinhart Aquarium; and Stan Minasian of EarthViews. Also most helpful were the California Academy of Sciences, the San Francisco Public Library, and researcher Shellei Addison.

It would be impossible for a layman, at least this lay-man, to attempt a project of this kind without relying heavily on the work of animal behaviorists, biologists, zoologists, photographers, and other intrepid field workers. That most of them were largely unaware of their contributions makes them no less appreciated. These include: Heather Angel, A. T. Band, Erwin A. Bauer, Fred Bavendam, Carol Beckwith, G. I. Bernard, Kevin Bond, John Brackenbury, Jane Burton, Larry Cameron, Densey Clyne, Bill Curtsinger, S. Dalton, Peter David, E. R. Degginger, Tui De Roy, Terry Domico, Alain Endewelt, Patrick Fagot, William E. Ferguson, Michael and Patricia Fogden, Ron Garrison, Philip Green, Crawford H. Greenewalt, Daniel Harper, Hal H. Harrison, Frants Hartmann, Robert W. Hernandez, C. M. Hladik, Eric Hosking, Jacana, Peter Johnson, M. P. Kahl, Takayoshi Kano, Alex Kerstitch, Tim Kimmel, Dwight Kuhn, Reinhard Künkel, Frank W. Lane, Frans Lanting, Phillip Lobel, G. Mazza, Tom McCalmon, Ron McGill, Skip Moody, Tom Myers, National Geo-graphic Society, Robbi Newman, Marie and Brian Pence, Feodor Pitcairn, Hans Reinhard, Marianne Riedman, Edward Rooks, L. Lee Rue, R. W. and E. A. Schreiber, Robert Sisson, Ron and Valerie Taylor, Wil-liam E. Townsend, Jr., M. Walker, Tom Walker, G. Zeisler, and Dale and Marion Zimmerman.

# BIBLIOGRAPHY

Arnold, Caroline. *Orangutan.* New York: Morrow Junior Books, 1990.

Bavendam, Fred. "Eye to Eye with the Giant Octopus." *National Geographic,* March 1991: 86–97.

Beckwith, Carol, and Marion Van Offelen. *Nomads of Niger.* New York: Abradale Press, Harry N. Abrams, 1993.

Brackenbury, John. *Insects Life Cycles and the Seasons.* U.K.: A Blanford Book, 1994.

Brooks, Bruce. *On the Wing.* New York: Charles Scribner's Sons, 1989.

Bull, John, and John Farrand, Jr. *The Audubon Society Field Guide to North American Birds, Eastern Region.* New York: Alfred A. Knopf, 1977.

Burn, Barbara. *North American Mammals.* New York: Bonanza Books, 1984.

Burton, Robert. *The Mating Game.* New York: Crown Publishers, 1976.

Bustard, Robert. *Sea Turtles: Natural History and Conservation.* New York: Taplinger Publishing Co., 1972.

Campbell, Bruce. *The Dictionary of Birds in Color.* New York: Viking Press, 1974.

Catton, Chris, and James Gray. *Sex in Nature.* New York and Oxford: Facts on File, 1985.

Chadwick, Douglas H. "Elephants— Out of Time, Out of Space." *National Geographic,* May 1991: 2–49.

Clutton-Brock, Juliet. *Cat.* New York: Alfred A. Knopf, 1991.

Cochran, Doris M. *Living Amphibians of the World.* Garden City, N.Y.: Doubleday & Co., 1967.

Cogger, Harold G., and Richard G. Zweifel, cons. eds. *Reptiles & Amphibians.* Smithmark Publishers, 1992.

*David Attenborough's The Trials of Life.* Video. 1991. Distributed by Time-Life Video, Alexandria, VA.

Department of the Navy: Bureau of Medicine and Surgery. "Poisonous Snakes of the World: A Manual for Use by U.S. Amphibious Forces." Washington D.C.: U.S. Government Printing Office, NAVMED P-5099.

Domico, Terry. *Kangaroos: The Marvelous Mob.* New York and Oxford: Facts on File, 1993.

Duplaix, Nicole. *World Guide to Mammals.* New York: Crown Publishers, 1976.

Emmel, Thomas C. *Butterflies.* Mallard Press, 1991.

*The Encyclopedia Britannica,* Fifteenth Edition. Chicago: The University of Chicago, Encyclopedia Britannica, Inc., 1985.

Evans, Michael Edwin Glyn. *The Life of Beetles.* New York: Hafner Press, 1975.

Farrand, Jr., John. *The Audubon Society Encyclopedia of Animal Life.* New York: Clarkson N. Potter, 1982.

Fenton, M. Brock. *Bats.* New York and Oxford: Facts on File, 1992.

Fischer-Nagel, Heiderose. *Life of the Butterfly.* Carolrhoda Books, 1987.

Fisher, Harvey I., and Mildred L. Fisher. *Wonders of the World of the Albatross.* New York: Dodd, Mead, 1974.

Forslund, Scott. "The Sex Life of Slugs." *Pacific Northwest,* May 1983: 32.

Galdikas, Biruté M. F. "Indonesia's Orangutans: Living with the Great Orange Apes." *National Geographic,* June 1980: 830–53.

Gamlin, Linda, and Gail Vines, eds. *The Evolution of Life.* New York: Oxford University Press, 1987.

Garth, John S. *California Butterflies.* Berkeley, Cal.: University of California Press, 1986.

Gentry, Roger L. "Seals and Their Kin." *National Geographic,* April 1987: 474–501.

Green, Philip. "The Satin Bowerbird: Australia's Feathered Playboy." *National Geographic,* December 1977: 864–70.

Greenewalt, Crawford H. "The Marvel-ous Hummingbird Rediscovered." *National Geographic,* July 1966: 98–101.

Grilhé, Gillette. *The Cat and Man.* New York: G. P. Putnam's Sons, 1974.

Grosvenor, Melville Bell, ed. *Wild Animals of North America.* Washington, D.C.: The National Geographic Society, 1960.

Grzimek, H. C. Bernhard, ed. *Grzimek's Animal Life Encyclopedia.* New York, Cincin-nati, Toronto, London, and Melbourne: Van Nostrand Reinhold, 1975.

Guggisberg, C.A.W. *Wild Cats of the World.* New York: Taplinger Publishing Company, 1975.

Hall, Alice J. "Man and Manatee: Can We Live Together?" *National Geographic,* September 1984: 400–418.

Halliday, Tim, ed. *Animal Behavior.* Norman, Okla.: University of Oklahoma Press, 1994.

Hammond, Michael. "The Incredible Pygmy Chimpanzee." *ZooNooz,* August 1986: 5.

Harrison, Colin James Oliver. *Bird Families of the World.* New York: Harry N. Abrams, 1978.

Harrison, Hal. H. *A Field Guide to the Birds' Nests—United States East of the Mississippi River.* New York: Houghton Mifflin, 1975.

Hayes, Jean. "The Golden Weaver." *Africana,* March 1969: 18–19.

Hornocker, Maurice G. "Learning to Live with Mountain Lions." *National Geographic,* July 1992: 52–65.

Howe, William H. *The Butterflies of North America.* Garden City, New York: Doubleday & Co., 1975.

Jackson, Rodney, and Darla Hillard. "Tracking the Elusive Snow Leopard." *National Geographic,* June 1986: 793–809.

Johnsgard, Paul A. *Hawks, Eagles and Falcons of North America.* Washington, D.C.: Smithsonian Institute, 1990.

Johnson, Sylvia A., photographs: Frans Lanting. *Albatrosses of Midway Island.* Minneapolis: Carolrhoda Books, 1990.

Johnson, Sylvia A. *Tree Frogs.* Lerner Publications Company, 1986.

Kano, Takayoshi. "The Bonobos' Peaceable Kingdom." *Natural History,* November 1990: 62–71.

Kohn, Bernice. *All Kinds of Seals.* New York: Random House, 1968.

Künkel, Reinhard. *Elephants.* Translated from the German by Ursula Korneitchouk. New York; Harry N. Abrams, 1989.

Laycock, George. *North American Wildlife.* New York: Exeter Books, 1983.

*Larousse Encyclopedia of Animal Life.* Middlesex, England: The Hamlyn Publishing Group, 1967.

León, Vicki. *Seals & Sea Lions: An Affectionate Portrait.* San Luis Obispo, Cal.: Blake Publishing, 1988.

Lewin, Roger. *Thread of Life.* Washington, D.C.: Smithsonian Books, 1982.

Linden, Eugene. "Bonobos—Chimpanzees with a Difference." *National Geographic,* March 1992: 46–53.

Line, Les, ed. *The Audubon Society Book of Marine Wildlife.* New York: Harry N. Abrams, 1980.

Line, Les, ed. *The Audubon Society Book of Wild Animals.* New York: Harry N. Abrams, 1977.

Linley, Mike. *Discovering Frogs and Toads.* New York: The Bookwright Press, 1986.

Löfgren, Lars. *Ocean Birds.* New York: Crescent Books, 1987.

Lopez, Barry. "An Elusive Cat." *Geo,* June 1981: 107–16.

MacClintock, Dorcas. *A Natural History of Zebras.* New York: Charles Scribner's Sons, 1976.

Mann, Charles A. "Macaws." *National Geographic,* January 1994: 118–40.

Marten, Michael, John May, and Rosemary Taylor. *Weird & Wonderful Wildlife.* San Francisco: Chronicle Books, 1982.

Milne, Lorus, and Margery Milne. *The Audubon Society Field Guide to North American Insects and Spiders.* New York: Alfred A. Knopf, 1980.

Milne, Lorus Johnson. *A Time To Be Born.* San Francisco: Sierra Club, 1982.

Minton, Sherman. *Venomous Reptiles.* New York: Charles Scribner's Sons, 1980.

*The Nature of Sex.* 6 programs. 1992. Genesis Films Productions Ltd. (U.K.)

*Our Amazing World of Nature.* New York: The Reader's Digest Association, 1969.

Owen, Denis. *Grasslands of Africa.* New York: The National Audubon Society, 1981.

Parker, Steve. *Fish.* New York: Alfred A. Knopf, 1990.

Riedman, Marianne. *Sea Otters.* Monterey, Cal.: Monterey Bay Aquarium, 1994.

Ross, Arnold, and William K. Emerson. *Wonders of Barnacles.* New York: Dodd, Mead & Company, 1974.

Ross, Edward S. "Mantids: The Praying Predators." *National Geographic,* February 1984: 268-79.

Rudloe, Anne, and Jack Rudloe. "Sea Turtles—In a Race for Survival." *National Geographic,* February 1994: 94–121.

Sadoway, Margaret Wheeler. *Owls: Hunters of the Night.* Lerner Publications Co., 1981.

Sagan, Dorian. *Garden of Microbial Delights: A Practical Guide to the Subvisible World.* Orlando, Fla.: Harcourt Brace Jovanovich, 1988.

Savage, Candace. *Grizzly Bears.* San Francisco: Sierra Club, 1990.

Schmidt, Karl P., and Robert F. Inger. *Living Reptiles of the World.* Garden City, N.Y.: Doubleday & Co., 1957.

Seidensticker, John, and Susan Lumpkin, cons. eds. *Great Cats.* Emmaus, PA.: Rodale Press, 1991.

Silcock, Lisa, ed. *The Rainforests.* San Francisco: Chronicle Books, 1989.

Silverstein, Alvin, and Virginia Silverstein. *Rabbits: All About Them.* New York: Lothrop, Lee and Shepard Co., 1973.

Slack, Gordy. "Fishes Whisper Sweet Somethings." *Pacific Discovery,* Winter 1992: 4.

Stebbins, Robert. *Amphibians of Western North America.* Berkeley and Los Angeles: University of California Press, 1962.

Stutz, Bruce D. "One Singular Sensation . . ." *Natural History,* October 1986: 106-7.

Terres, John. *The Audubon Society Encyclopedia of North American Birds.* New York: Alfred A. Knopf, 1980.

Vecchio, Tony. "Hamadryas Hierarchies." *Riverbanks,* July–August 1985: 2–7.

Voss, Gil. "The Octopus." *National Geographic,* December 1971: 775–99.

Wallace, Robert Ardell. *How They Do It.* New York: William Morrow and Co., 1980.

Weyer, Edward M. *Strangest Creatures on Earth.* New York: Sheridan House, 1953.

Whitaker, Jr., John O. *The Audubon Society Field Guide to North American Mammals.* New York: Alfred A. Knopf, 1980.

Wilson, Roberta, and James Q. Wilson. *Watching Fishes: Life and Behavior on Coral Reefs.* New York: Harper & Row Publishers, 1985.

Whitfield, Dr. Philip. *MacMillan Illustrated Animal Encyclopedia.* New York: MacMillan Publishing Co., 1984.

Wolfe, Art. *Bears: Their Life and Behavior.* New York: Crown Publishers, 1993.

# C O L O P H O N

Most of the *illustrations for this book were drawn with graphite pencil, augmented by watercolors, on Arches watercolor paper. The others were drawn with Derwent watercolor pencils on Crescent black drawing paper.*

*The book was designed by the author, assisted by Tom Morgan of Blue Design. Jennifer Petersen set the type. The typefaces are digitized versions of Centaur, originally designed by Bruce Rogers, and its companion italic, Arrighi, designed by Frederic Warde.*

*The book was printed by offset lithography by Mondadori A.M.E. Publishing Ltd., Italy, from separations of the original art. The paper is Gardamatt Brillante Demimatt Coated and the binding cloth is Imitlin Embossed.*